SMART WORDS READER

BUILT FOR SPEED

Vicky Willows

SCHOLASTIC INC.

What are SMART WORDS?

Smart Words are frequently used words that are critical to understanding concepts taught in the classroom. The more Smart Words a child knows, the more easily he or she will grasp important curriculum concepts. Smart Words Readers introduce these key words in a fun and motivational format while developing important literacy skills. Each new word is highlighted, defined in context, and reviewed. Engaging activities at the end of each chapter allow readers to practice the words they have learned.

No part of this publication may be reproduced, stored in a retrieval system, or transmitted in any form or by any means, electronic, mechanical, photocopying, recording, or otherwise, without written permission of the publisher. For information regarding permission, write to Scholastic Inc., Attention: Permissions Department, 557 Broadway, New York, NY 10012.

ISBN 978-0-545-93909-6

Copyright © 2016 by Scholastic Inc.

Photos ©: cover: Mark Newman/Getty Images; 1: Janina Kubik/Dreamstime; 2-3 background: TongRo Images/Thinkstock; 3: Glenn Price/Shutterstock, Inc.; 4-5 background: Nature Picture Library/Alamy Images; 5 car inset: Mark Evans/iStockphoto; 5 cheetah inset: Gallo Images-Heinrich van den Berg/Getty Images; 6-7 spread: Rajeev Doshi/Getty Images; 7 inset: Kandfoto/iStockphoto; 8-9 background: age fotostock/Alamy Images; 9 inset: robertharding/Alamy Images; 10-11 background: Biosphoto/Superstock, Inc.; 10 inset: Nigel Cattlin/Science Source; 11 inset: Konrad Wothe/Minden Pictures; 12 main: Nature Picture Library/Alamy Images; 12 soil/sand texture and textures throughout: CG Textures; 13: Jasperdebeer/Dreamstime; 14-15 background: Sergey Kohl/Shutterstock, Inc.; 14 inset: Panaiotidi/Shutterstock, Inc.; 16-17 background: Jim Zipp/Ardea; 16 inset: Chris Hill/Shutterstock, Inc.; 17 inset: Stuart Price; 18: paula french/Shutterstock, Inc.; 19: Sekar B/Shutterstock, Inc.; 20 damselfly: dennisvdw/iStockphoto; 21: FLPA/David Tipling/age fotostock; 22-23 background: Doug Perrine/Alamy Images; 22 inset: holbox/Shutterstock, Inc.; 24-25: suriyasilsaksom/iStockphoto; 26-27 background: robertharding/Alamy Images; 27 inset: Designua/Shutterstock, Inc.; 27 inset marlin: seamartini/iStockphoto; 28 reef fish: Fred Bavendam/Minden Pictures; 29: WaterFrame/Alamy Images; 30-31: ssuaphoto/iStockphoto.

All rights reserved. Published by Scholastic Inc., *Publishers since 1920*. SCHOLASTIC, SMART WORDS READER, and associated logos are trademarks and/or registered trademarks of Scholastic Inc.

The publisher does not have any control over and does not assume any responsibility for author or third-party websites or their content.

10 9 8 7 6 5 4 3 2 1   16 17 18 19 20

Printed in the U.S.A.   40
First printing, September 2016

Designed by Marissa Asuncion

Table of Contents

Chapter 1: Fastest Animals on Land......... 4
   Use Your Smart Words........................ 12

Chapter 2: Fastest Animals in the Air..... 14
   Use Your Smart Words........................ 20

Chapter 3: Fastest Animals in the Sea... 22
   Use Your Smart Words........................ 28

Glossary 30

Index, Answer Key.. 32

Chapter 1

Fastest Animals on Land

Have you ever raced your friend to the playground? How fast can you run? If you know how far you run and have someone time you with a stopwatch, you can know your **speed**. Speed is the rate at which an object covers a certain distance over time.

If you raced a friend around the block, you would change direction at each corner. Then your **velocity** would change. Velocity describes speed in a particular direction.

The world's fastest car can **accelerate** from 0 to 60 miles per hour (97 kilometers per hour) in about 2.5 seconds. That's fast! Acceleration describes the change of an object's velocity over time.

Earth's fastest land animal, the cheetah, is about the same. It can accelerate from 0 to 60 miles per hour (97 kilometers per hour) in under 3 seconds. However, it can only maintain those speeds for a few minutes.

SMART WORDS

speed the rate at which an object covers a certain distance over time

velocity speed in a particular direction

acceleration the change in an object's velocity over time

Cheetah: Made for Speed

Both the cheetah and a sports car are built for speed. Being the fastest land animal isn't just about having the biggest muscles or the longest legs. Like a fast car, it must be able to burn fuel, which is used for energy. Cars use gasoline for energy, and living things use food.

Fuel, or food, in a living thing is stored in cells, the smallest unit of any living thing. In order for any type of fuel to be used for energy, it must be broken down. This process requires oxygen, which animals get from breathing.

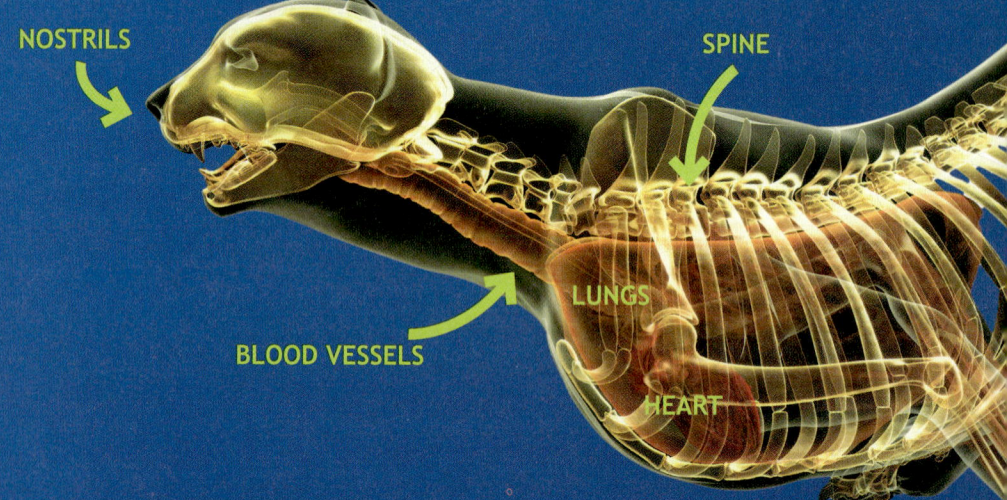

NOSTRILS

SPINE

BLOOD VESSELS

LUNGS

HEART

The faster something runs, the more energy it needs. This means more oxygen has to get to the muscle cells. This is the job of the heart and blood vessels.

Cheetah: Fast Facts

- Top speed: about 75 mph (121 kph)
- Large nostrils take in extra oxygen
- Extra-large lungs take in more oxygen
- Extra-large heart pumps oxygen-rich blood to muscle cells
- Long, slender legs and long tail balance and steer
- Flexible spine allows for long strides

With large nostrils, and an extra-large heart and lungs, the body of a cheetah is built to take in more oxygen for it to reach muscle cells efficiently.

The cheetah's body is built for speed in other ways, too. Besides having long, slender legs, the cheetah has a flexible spine. The backbone curves in a way that allows the cheetah's legs to swing out for an incredible stride.

SMART WORDS

cell the smallest unit of any living thing
spine the backbone

Predator and Prey: The Race for Survival

With the ability to accelerate to top speeds quickly, the cheetah is a successful predator. A predator is an animal that kills and eats other animals to survive. However, a cheetah must outrun its prey, the animal being hunted, before it runs out of speed.

Imagine that a cheetah is released into North America, where Earth's second-fastest land animal, the pronghorn, lives. The pronghorn resembles an antelope and spends its days grazing.

As the chase begins, the cheetah is clearly the faster runner. The pronghorn's top recorded speed is only 60 miles per hour (97 kilometers per hour). Within a few minutes, the cheetah runs out of energy and gives up the chase.

The pronghorn keeps going strong. It can maintain a fast speed for a longer period of time than the cheetah. It often outruns its North American predators, such as bobcats and coyotes.

What happens if both the prey and the predator are equally fast runners? Then strength usually wins. Lions run at about the same speed as wildebeests—both considered among Earth's top runners. The lion usually wins, as lions have powerful claws and teeth. However, there are instances of the prey besting the predator's strength.

a lioness chasing after a wildebeest

SMART WORDS

predator an animal that kills and eats other animals to survive

prey an animal that is killed and eaten by another animal

Mighty Mites

When comparing speed, sometimes size matters. It's not really fair to compare the speed of a small bug and a cheetah because of the difference in size. To eliminate that problem, speed is sometimes compared using body length. This measures speed relative to size.

Using this method, the fastest animal on Earth is a tiny arachnid called a mite. An arachnid is an animal with a two-part body, eight segmented legs, and a hard outer skeleton. Other arachnids include spiders and ticks.

Mite: Fast Facts

Name: *Paratarsotomus macropalpis*
Weight: 0.07 oz (2 g)
Length: up to 0.027 in (0.7 mm)
Habitat: Southern California, rocky areas
Speed: 322 body lengths per second

To compare, a mite can run up to 322 times its body length per second. A cheetah can only run about 16 times its body length per second at full speed. The fastest recorded human on Earth has a top speed of only 6 body lengths per second. The amazing mite can run at a speed equivalent to a human running at about 1,300 miles per hour (2,100 kilometers per hour)!

Other tiny creatures also take top spots for speed. These include **insects**, which have three body parts, six legs, a hard outer body covering, and usually one or two pairs of wings. The Australian tiger beetle comes in second by running at 171 body lengths per second. The relative human speed would be 240 miles per hour (386 kilometers per hour). That's faster than a race car!

the American cockroach can run at speeds equivalent to a human running 210 miles per hour (338 kilometers per hour)

SMART WORDS

arachnid an animal with eight segmented legs, a body divided into two parts, and a hard outer skeleton

insect a small animal with six legs, a body divided into three parts, a hard outer covering, and usually one or two pairs of wings

Use your SMART WORDS

Match each description with the correct Smart Word.

> speed   arachnid   prey   insect   spine
> predator   velocity   acceleration   cell

1. the change in an object's speed and direction over time
2. the smallest unit of a living thing
3. a small animal with six legs, a body divided into three parts, a hard outer covering, and usually one or two pairs of wings
4. the rate at which an object covers a certain distance
5. an animal that kills and eats other animals to survive
6. an animal with eight segmented legs, a body divided into two parts, and a hard outer skeleton
7. speed in a particular direction
8. an animal that is killed and eaten by another animal
9. the backbone found in some animals

Answers on page 32

Talk Like a Scientist

Use your Smart Words to talk about your favorite fast animal. Is it a predator or prey? Is it big or small?

SMART FACTS

That's Amazing!

A flea can jump a distance 200 times its body length. This feat takes 20 times the acceleration needed for a space shuttle to lift off.

Did You Know?

Mites are not only fast. They can run on concrete up to 140°F (60°C). That's almost hot enough to fry an egg on a sidewalk!

Good to Know

Although cheetahs are fast, they are not tough. They are lightweight and have blunt claws. If a more aggressive animal tries to take its food, the cheetah will usually give it up.

Chapter 2

Fastest Animals in the Air

Flight is our fastest form of transportation. Much of what we have learned about flight comes from the study of birds. When scientists apply lessons from animals in nature to the design of technology, it is called **biomimetics**.

Most birds have amazing **adaptations** for flight. An adaptation is a trait that enables a living thing to become better equipped to survive in its environment. The bones of a bird are hollow, making them strong but lightweight. They also have an enlarged breastbone to which flight muscles are attached.

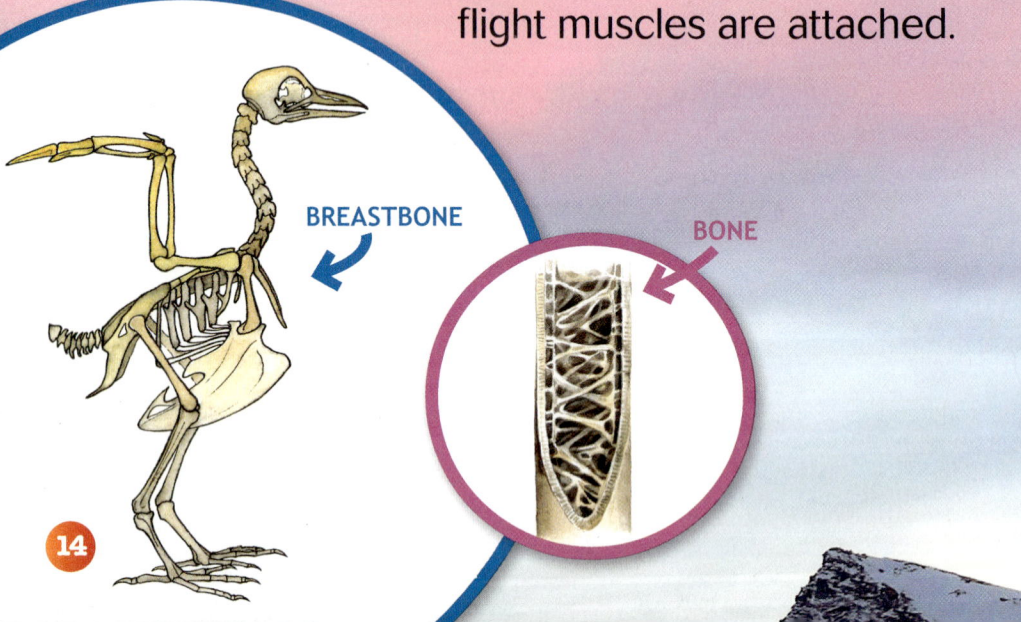

BREASTBONE

BONE

Like airplane wings, bird wings have an airfoil design. This simply means that the upper curve of the wing makes air travel over the top surface faster. This reduces the pressure on top of the wing. When this happens, the pressure below the wing is greater, pushing the bird into flight.

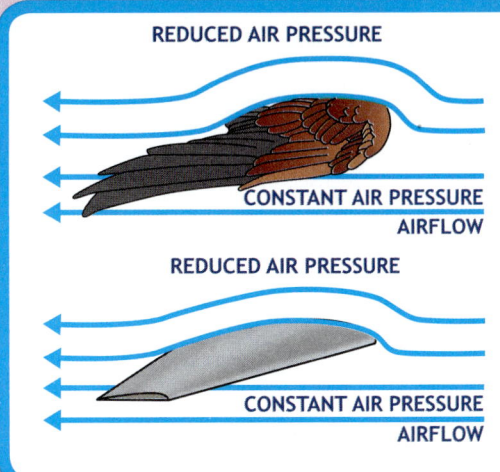

There are many different wing shapes and feathers. Each meets the needs of individual birds. Not all birds are built for speed. However, those that *are* break all the speed records in the animal kingdom

peregrine falcon

SMART WORDS

biomimetics the application of observations from animals in nature to the design of technology

adaptation a trait that enables a living thing to become better equipped to survive in its environment

airfoil a structure designed with curved surfaces to provide lift in flight

Peregrine Falcon: Earth's Fastest Animal

Peregrine falcons are predators. They can spot prey from at least 1 mile (1.6 kilometers) away. Once the prey is targeted, a falcon will drop into a near-vertical dive that can top 200 miles per hour (322 kilometers per hour).

As it goes into a dive, the bird folds its wings in tightly about its body. This makes the falcon more **aerodynamic**, allowing it to move even faster through the air. Just before reaching the prey, the falcon pulls out of the dive and prepares its **talons**. Another animal that is hit by these sharp claws is instantly killed.

Peregrine Falcon: Fast Facts

Size: 14–19 inches (36–48 centimeters)
Wingspan: 3.3–3.6 feet (1–1.1 meter)
Average flight speed: 40–55 miles per hour (64–89 kph)
Habitat: all continents except Antarctica
Prey: other birds, small mammals

The peregrine falcon is, indeed, Earth's fastest animal. However, this is only when the bird is in a full dive. When cruising in level flight, the falcon flies at about 40–55 miles per hour (64–89 kilometers per hour).

The white-throated needletail is found mostly in Australia. These birds have top speeds of 105 miles per hour (169 kilometers per hour). These speeds are reached when not in a full dive but cruising along in normal flight.

white-throated needletail

SMART WORDS

aerodynamic the way air moves around things; designed to move through air

talon sharp claw, particularly one belonging to a bird of prey

Flightless Runners

For many years, there was a cartoon featuring a coyote and a roadrunner. In each episode, the coyote was wildly unsuccessful in catching his prey, a flightless bird called a roadrunner. But could a coyote catch a roadrunner in real life?

Flightless birds are incredibly fast. The fastest is the ostrich. Using long, powerful legs, an ostrich can run at speeds up to 43 miles per hour (69 kilometers per hour). Although they can't lift off the ground, their wings help them change direction as they run.

The top speed of a coyote is also 43 miles per hour (69 kilometers per hour). However, an ostrich stands at about 8 feet (2 meters) tall and has extremely sharp claws on each of its two toes. The coyote would probably have to go home empty-handed . . . if he is a smart coyote.

ostrich

roadrunner

How do you think the roadrunner, the second-fastest flightless bird, would do against the coyote? Roadrunners can run as fast as 26 miles per hour (42 kilometers per hour). They can easily run down prey such as small reptiles and rodents, but they are quite slower than the coyote.

Roadrunners are also much smaller than the predator. They are about 18–22 inches (46–56 centimeters) in length. However, they have a few tricks up their feathers. Instead of running in a straight line, they zigzag and hide in low shrubs found in their habitat, or natural home. But it seems that a real coyote just might have a chance.

SMART WORD

habitat an organism's natural home

Use your SMART WORDS

Match each description with the correct Smart Word.

> habitat   aerodynamic   talon
> biomimetics   adaptation   airfoil

1. the way air moves around things, or designed to move through air
2. the application of observations from animals in nature to the design of technology
3. sharp claw, particularly one found on birds of prey
4. a trait that enables a living thing to become better equipped to survive in its environment
5. a structure designed with curved surfaces to provide lift in flight
6. an organism's natural home

Answers on page 32

Talk Like a Scientist

Use your Smart Words to describe other animals that might be used as models for the design of new technology.

SMART FACTS

That's Amazing!

The peregrine falcon has been used in the sport of falconry—hunting with a trained bird of prey—for over 3,000 years.

Did You Know?

It may not be able to fly quickly, but the gentoo penguin is the fastest swimming bird. It can reach speeds of 22 miles per hour (35 kilometers per hour).

Good to Know

Ostriches don't really bury their heads in the sand as many people think. Animal experts believe they lie down on the ground with their neck outstretched to hide when they feel in danger.

Chapter 3

Fastest Animals in the Sea

Finally, we head off to find the speediest animals in the sea. It's more challenging for these animals to break records for speed because water is about 750 times denser than air.

Density is a measure of how much matter is in a given space. You might describe it as how "thick" something is. The effect of density is that water is more difficult to move through than air. Think about how difficult it is to walk through waist-high water compared to walking on dry land!

bluefin tuna

The bodies of fish are perfect for moving through water. Their bodies are thin and pointed at both ends. Fish also have **fins**, which are wing-like or paddle-like limbs that are used for swimming and balance.

Fish breathe through gills, which are the respiratory organs of some aquatic animals. Gills allow fish to obtain oxygen that is dissolved in the water. However, some sea animals, such as whales, are not fish. They are mammals and have lungs, like you.

sailfish swimming

SMART WORDS

density a measure of how much matter is in a given space

fin a wing-like or paddle-like structure that is used for swimming and balance

gills respiratory organs for some aquatic animals

Sailfish: Fastest Animal in the Sea

You just have to look at a sailfish to see that is it a lean, mean swimming machine. It can reach speeds of 68 miles per hour (109 kilometers per hour), which is not that much slower than the cheetah, at about 75 miles per hour (121 kilometers per hour).

The "sail" on the sailfish is actually a dorsal fin. Dorsal always refers to the upper side of an animal. The dorsal fin on a sailfish stretches nearly the length of its body. It is often kept down to make swimming easier. When the fish gets excited or is threatened, the sail is raised.

sailfish

Sailfish are found in the Atlantic and Pacific Oceans. They can grow up to 11 feet (3 meters) long and weigh between 120–220 pounds (54–100 kilograms). They will sometimes use their dorsal fin to herd together small fish that they like to eat. In addition to their large dorsal fin, sailfish have a long bill shaped like a spear that they use to catch prey.

SMART WORD

dorsal upper side of an animal

Marlin: Top of the Chain

Like the sailfish, the marlin has a long pointed bill and large dorsal fin. It holds the record as the second-fastest fish in the sea, reaching speeds of about 50 miles per hour (80 kilometers per hour). The marlin is also one of the biggest fish in the world. The average length is about 11 feet (3 meters) and their average weight is about 200–400 pounds (91–181 kilograms).

The marlin is a dramatic hunter. It will slash through a school of fish with its spear-like bill and then circle back to eat those that were wounded.

blue marlin catching prey

As predators, marlins are one of the big fish at the top of the ocean **food chain**. A food chain shows the flow of energy in an ecosystem, such as the ocean. Energy for all living things comes directly from the sun. In the ocean, tiny organisms at the surface of the water use energy from the sun to make food. Many fish and mammals in the sea eat these tiny organisms. Often, bigger fish and organisms eat the smaller fish, until you get to the top level.

Whether on land, in the air, or in the sea, the speed of an animal will determine if it will have a meal or avoid being a meal. Most often, the faster animal wins the predator-prey battle!

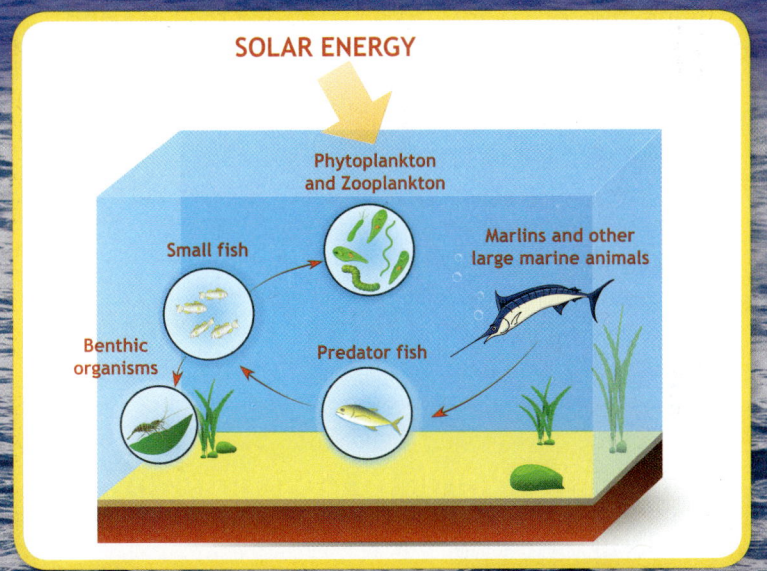

SMART WORD

food chain a model used to show the flow of energy through an ecosystem

Match each description with the correct Smart Word.

1. I am a measure of how much matter is in a given space.
2. I am a model of how energy flows through an ecosystem.
3. I am the respiratory organs for some aquatic animals.
4. I am a wing-like or paddle-like structure that is used by fish for swimming.
5. I am the upper side of an animal.

Answers on page 32

Talk Like a Scientist

Use your Smart Words to describe how fish move through the sea.

SMART FACTS

That's Amazing!

Flying fish can leap out of the water up to 4 feet (1 meter) in the air. They can then glide for about 655 feet (200 meters) before going back into the water.

Did You Know?

Sailfish, swordfish, and marlins would get ticketed on US highways! They swim faster than most highway speed limits.

Good to Know

The sailfish might be working out! It has huge amounts of muscle that help it accelerate quickly.

Glossary

acceleration the change in an object's velocity over time

adaptation a trait that enables a living thing to become better equipped to survive in its environment

aerodynamic the way air moves around things; designed to move through air

airfoil a structure designed with curved surfaces to provide lift in flight

arachnid an animal with eight segmented legs, a body divided into two parts, and a hard outer skeleton

biomimetics the application of observations from animals in nature to the design of technology

cell the smallest unit of any living thing

density a measure of how much matter is in a given space

dorsal upper side of an animal

fin a wing-like or paddle-like structure that is used for swimming and balance

food chain a model used to show the flow of energy through an ecosystem

gills respiratory organs for some aquatic animals

habitat an organism's natural home

insect a small animal with six legs, a body divided into three parts, a hard outer covering, and usually one or two pairs of wings

predator an animal that kills and eats other animals to survive

prey an animal that is killed and eaten by another animal

speed the rate at which an object covers a certain distance over time

spine the backbone

talon sharp claw, particularly one belonging to a bird of prey

velocity speed in a particular direction

Index

acceleration 5	food chain 27
adaptation 14, 15	gills 23
aerodynamic 16, 17	habitat 19
airfoil 15	insect 11
arachnid 10, 11	predator 8, 9
biomimetics 14, 15	prey 8, 9
cell 6, 7	speed 4, 5
density 22, 23	spine 7
dorsal 24, 25	talon 16, 17
fin 22, 23	velocity 4, 5

SMART WORDS Answer Key

Page 12
1. acceleration 2. cell 3. insect 4. speed 5. predator 6. arachnid 7. velocity 8. prey 9. spine

Page 20
1. aerodynamic 2. biomimetics 3. talon 4. adaptation 5. airfoil 6. habitat

Page 28
1. density 2. food chain 3. gills 4. fin 5. dorsal